littlepinkhouses

Little Pink Houses
Copyright © 2019 by Abbe Quarles

abbequarles@gmail.com
AQ Collective Press
aqcollective.com

Cover and Interior Design by Lance Buckley
www.lancebuckley.com

Available in Paperback, Kindle, and eBook formats.

Paperback ISBN: 978-1-7343556-0-4
eBook ISBN: 978-1-7343556-1-1

Buying | Renovating | Selling

littlepinkhouses

Five Assignments
for Putting All of
the Pieces Together

abbequarles

For Griffin, Ford, Lake,
Worth, Ophelia,
...and Steve

CONTENTS

INTRODUCTION..1

PERSONAL FINANCE...5
   Fantasy Meets Reality...5
   Hindsight..7
   Assignment 1...9
   Have a Clear Understanding of Your Expenses.....9
   The Meeting...11

PLAN..14
   Little Timmy...14
   Couple's Compromise.......................................16
   Assignment 2...18
   Organize a Strategic Plan..................................18

PREPARE...23
   Honey, I Have a Great Idea!...............................23
   Be Ready to Make an Offer................................27
   Assignment 3A...28
   Manage the Market...28
   Assignment 3B...29
   Manage Money...29
   Discussions..30

PRIORITIZE..33

    Put All of the Pieces Together...................................34

    Pick Your Battles...35

    Assignment 4..36

    Educate, Evaluate, and Edit.......................................36

    Educate:..36

    Prioritize, Prioritize, Prioritize................................37

    Evaluate and Edit:..38

PERSONALIZE..42

    A Few Quick Questions..43

    Conceptualize...45

    The Mantra...46

    Assignment 5..48

    Stage and Style...48

PROSPER...52

AFTERWORD...56

HOMES WORK..59

ABOUT THE AUTHOR..73

INTRODUCTION

Are you thinking about buying your first home? Are you trying to decide whether to sell your current home? Are you considering a complete home renovation or remodeling an existing home?

If so, you are beginning the initial phase of educating yourself on the adventures of home-buying, renovating, and selling. This process can be overwhelming and endlessly challenging, but with the proper amount of preparation, aligning of expectations, and establishing your goals, you can develop a plan of action with specific guidelines for the best possible outcome.

This experience will ignite your creativity and challenge your mathematical skills, as it is very much both a right-brain and left-brain activity. There are hundreds of questions to answer: from expenditures and spreadsheets, to budgeting and scheduling, to design techniques, along with curating your own personal artistry and style.

There are multiple aspects of the home development

process, including thoroughly tracking your household expenses, navigating and understanding the language of mortgages and loans, developing strategic goals, and educating yourself on your local real estate market.

The list of questions you will be expected to answer can be lengthy and needed in a timely manner, which without proper preparation can cause the process to become overwhelming, causing undo delays. We have all heard the phrase "time is money." This expression definitely applies to home-buying, renovating, and selling.

Contracts can expire, interest rates can change, opportunities can be lost, and scheduling delays can be exorbitantly expensive. Using scheduling as an example, not being prepared to make decisions in a timely manner can result in product and material orders being delayed, vendors and subcontract laborers having to be rescheduled, increased interest payments on your home improvement loan, along with additional rent and mortgage payments on your current home. You do not want to find yourself involved in this chain of events.

That is why I have written this book. My hope is that the included list of building blocks will give you a firm foundation moving forward. I like to call these building blocks **HOMES** work assignments. There are five assignments in developing a solid plan:

1. Have a Clear Understanding of Your Expenses
2. Organize a Strategic Plan
3. Manage Money and Marketability
4. Educate, Evaluate, and Edit
5. Stage and Style

If you follow these explorations, along with the assignments at the end of this book, you will not only become familiar with the basic processes needed for an optimum experience, but you will also seamlessly work through each building block in a step-by-step manner as you conquer your goals. Carefully focusing on each block will help you establish your own personalized vision.

I've worked with many couples over the past twenty years and witnessed very different personalities, lifestyles, goals, and motivations. I have also worked with people of all ages and from many different socioeconomic backgrounds. I've seen couples who were clearly on the same page and others ready to wage their own version of *The War of the Roses*. I truly believe that if you arm yourself with the tools—information, careful planning, well-defined goals, and a journey into self-discovery—you can achieve a solid foundation to hammer through the home development process.

This information is based on years of experience working with clients, as well as buying, renovating, and selling many of my own projects. I have been a part of numerous projects from mid-century modern to classic Georgian. I have completed the educational requirements and examination for the Tennessee Board for Licensing Contractors, in addition to the educational requirements for obtaining my real estate license, and most importantly, the tremendous value of hands-on exposure in the field. I've had great success in real estate, and I'm a survivor of many things, one of them being the Great Recession and its 2008-2009 real estate crisis. I have learned that developing a firm foundation through education, preparation, and embracing the important skill of communication is both your best defense and your best offense. After all, you wouldn't buy a house, remodel a house, or build a house that wasn't engineered and constructed on a rock-solid foundation.

I hope that you will use the information in this book as your own personally constructed map to guide you and your family on your journey in pursuit of the great American dream of home ownership.

PERSONAL FINANCE
Have a Clear Understanding of Your Expenses

D early beloved, we are gathered here today to lovingly support Johnny and Susie, or for the sake of modernization, let's call them Lake and Lily. They are planning to purchase their ideal dream home, one that will be adorned with a perfectly painted white picket fence. They will start a family, have 2.5 children, and live happily ever after.

Their expressions are dreamy: "It's going to be beautiful! It will have a front porch swing, boast hundred-year-old live oaks, and bloom buttercups in early spring. First, we will rescue a show dog, and in two years, we'll have twins—a boy and girl. We'll build a roaring fire in our wood-burning fireplace each winter and have summer picnics in the backyard. And then—"

Uh, oh.

Lake? Lily? Have you done your HOMES work?

FANTASY MEETS REALITY

This is where the rubber meets the road or, should we say,

fantasy meets reality. Home-buying strategies, particularly the emotional side, are not part of the national or state standard high school curriculum, or at least they weren't where I went to school. Owning a home is part of the American dream, a part of the natural progression of life: "First comes love, then comes marriage, then comes pushing…the bank account?"

Oops!

Well, let's face it—some significant facts have been conveniently left out of the cute little song we sang on the playground, and the facts are not pretty.

On average, more than 50 percent of marriages end in divorce—that's one legal separation every thirteen seconds. Couples with little to no assets at the beginning of the third year of marriage are at an increased risk of relationship failure, which means financial issues are a major contributor. Considering that a home is likely the most expensive purchase a couple will make in their lifetime, they should realize the impact home-buying can have on their marriage.

We all want to create a vision for the future of making memories by celebrating holidays, spending time with friends, and possibly raising a family. This can be very overwhelming. Home-buying involves a complex set of factors to contemplate. While there is no perfect equation for purchasing your own home, the best ammunition is awareness and planning. Whether you are a first-time home buyer or planning to upgrade, it's time to go back to school to gain education on the financial and emotional aspects of selecting a home.

I have created five assignments to use as a guide. In these assignments, you will complete a step-by-step plan to explore and learn the many aspects of the home development journey by highlighting the infinite possibilities for you to consider: from lifestyles and location, to money and the real estate market, and other unanswered questions that many individuals and couples haven't explored. Entering into home-buying adventures in a forward-thinking manner should result in a positive and successful outcome.

HINDSIGHT

Unfortunately, I have witnessed many couples learn the hard way—through hindsight. While twenty-twenty vision does provide great clarity, no one needs to have a master's degree from the school of hard knocks. After all, Lake and Lily and many others are investing their hearts, minds, souls, time, and life savings into this level of commitment. So, taking the time to pause, process, and learn is essential in developing and customizing a personal plan.

First things first, you must have a clear and thorough understanding of any and all expenses. I can hear you now…

"Oh, I know exactly how much I spend."

"I shop at the local consignment stores."

"I only order out once a week."
"I only get my nails done once a month."
"I don't go out with the boys that often."
But do you REALLY know?

ASSIGNMENT 1

Have a Clear Understanding of Your Expenses

This lesson is a very personal journey into one's private financial life. It can create conflict in relationships, negatively affect physical health, illicit worry and fear, as well as take us on a wonderful journey into freedom and prosperity.

Having a detailed snapshot of ALL expenditures is the first step. Knowledge is power and will provide you with ammunition moving forward. I recommend investing in an expense-tracking software program. It's a minimal cost that will give you a lot of bang for your buck. QuickBooks or Quicken are my personal favorite accounting software packages developed by Intuit. They are user-friendly, and they link to your bank and credit card accounts, downloading all of your financial activity.

You will need to establish several categories in order to target, prioritize, and clearly understand your monthly expenses.

SUGGESTED CATEGORIES:

- Debt: apartment rent, car payments, student loans, credit cards, and any other legal obligations
- Tax obligations: income tax, and if you're self-employed, franchise and excise tax, schedules C or S. Consult with your

preferred accountant to gain an understanding of a potential yearly amount.

- Utilities: gas, electric, water, garbage, Wi-Fi, cable, streaming subscriptions, cell phones
- Insurance: renter's insurance, auto insurance, health insurance, life insurance
- Medical: co-pays on doctor visits and prescriptions
- Auto: gas, oil changes, maintenance
- Food, beverage, and entertainment: talk about a rude awakening—this one will surprise you at the end of the month.
- Travel: annual vacations, family visits, unreimbursed work-related travel
- Charitable contributions: church tithing, nonprofit donations
- Discretionary: haircuts, mani/pedis, gym memberships, gifts, cash

Warning! This exercise does have its challenges. It's tempting to play the blame game.

"I'm a better saver than you."

"You're the spender."

"I could drive any old car."

"You should do it my way."

Stop! Don't go there. Try thinking of it as a new diet plan. A financial diet. Take on the mind-set, "We're going to form an alliance. We're going to encourage each other. We're going to challenge and support each other."

THE MEETING

Communicating about finances is never easy. Several years ago, I worked with a couple who spoke one simple word that brought me many years of benefit when it comes to communicating.

That word was "meeting."

They had an admirable philosophy and commitment to routinely discussing their finances by calling it a meeting. They scheduled a specific time each week to discuss and evaluate that week's expenditures. By looking at finances openly, objectively, strategically, and intelligently, they eliminated the emotions. By focusing on spending as if it were a line item on a spreadsheet in a business arrangement, they defused any drama and potential marital trauma.

Genius!

The meetings, which were scheduled each Friday at 3:00 p.m., were held in a proper business-like setting, whether at her office or their kitchen table. They met with all the necessary financial paperwork, spreadsheets, receipts, and other pertinent information. The meetings lasted approximately an hour, and by the end of their collaboration, they had a clear assessment of the prior week's expenditures and a set of goals moving into the next week. I highly recommend this method. It works—I've seen it firsthand, and I've tried it myself.

New adventures do have challenges, but once you have

summited this learning curve, it will bring you a lifetime of skill, awareness, direction, and a new level of communicating—and hopefully a few extra dollars in the bank.

So, after ninety days of carefully evaluating monthly expenditures and successfully committing to routine financial meetings, you are now ready to move on to the next assignment.

Don't worry—it gets more fun!

"Real estate cannot be lost or stolen, nor can it be carried away. Purchased with common sense, paid for in full, and managed with reasonable care, it is about the safest investment in the world."

– FRANKLIN D. ROOSEVELT

PLAN

Organize a Strategic Plan

A ll right, all right, all right!
Now that Lake and Lily have successfully navigated ninety days of evaluating and adjusting their expenses and have developed open lines of communication about a potentially volatile subject, they're ready to start the wonderfully exciting process of selecting their first home together.

LITTLE TIMMY

Lily, with great passion and enthusiasm, says, "I can't wait to tour the new high-rise near the business district. The area is filled with trendy boutiques, book stores, and coffee shops. We could walk everywhere! Maybe even get a couple of bikes. Did you see the recently installed bike lanes? An urban lifestyle could be so much fun!"

"Urban lifestyle?" remarks Lake. "I thought we would visit traditional homes in the suburbs, maybe a bungalow or a Georgian—a classic style home with a nice backyard where I could have weekend barbecues and someday throw a football with little Timmy. We could even consider planning for an addition in ten years or so, as our family grows. We'll definitely need to check out the school zones, too."

"Little Timmy? Ten years? I thought we would enjoy the city life—you know, for the next three to five years."

"Move again??"

Uh, oh. Houston, we have a problem, or should I say, "Hooterville" for those of you familiar with the 1960s sitcom *Green Acres*. The opening song highlights the conflict between Lisa and Oliver Wendell Douglass, and calamity ensues. The lyrics express the couple's differences in choosing a manner of living. Oliver proudly sings, "Farm living is the life for me," while Lisa responds, "Darling, I love you, but give me Park Avenue."

Time for more HOMES work.

In the situation comedy, Lisa reluctantly forgoes the glamorous life in her luxury Manhattan penthouse to become a farmer's wife. Oliver, a successful New York City attorney, longs for a more modest lifestyle with quiet surroundings in a simple rural setting. Together they pack up their belongings, move to a small town to begin restoring a dilapidated shack, and settle into their new lifestyle.

Talk about a fixer-upper. Their house boasted of an outdoor shower, insufficient electrical power, a telephone that required climbing a utility pole, an intrusive ranch hand, a pig that understood English, and a list of quirky townspeople, all of which created daily frustrations while they acclimated to their new surroundings.

Oliver and Lisa are not exactly the best example of a couple's compromise. Something tells me that if Lake and Lily completely acquiesce to the other's expectations—unlike the sitcom—hilarity and humorous quips are not going to be a part of their daily routine.

COUPLE'S COMPROMISE

A couple's compromise is key to developing an understanding of each other's wants and needs when choosing a home. By evaluating their personal styles and understanding lifestyle differences, they can successfully create a definitive plan for the future. Without doing so, couples can easily slip into a merry-go-round of tensions and frustration with resentment soon setting in, taking the fun and enthusiasm out of what should be an exciting journey through the adventure of selecting and curating a home.

A compromise is "an ability to listen to two sides in a dispute and devise a compromise acceptable to both." (*The Oxford Dictionary of Difficult Words*)

I worked with a couple I will call Peter and Penelope. Peter and Penelope had very different views in what they wanted in a home. We set out to find a house with good bones that could support the level of remodeling that they could afford. They were both very creative and wanted to customize their own space. It became apparent that they each had very different visions. They couldn't agree on location, amenities, floor plans, or even paint colors. A basic concept didn't exist. The process became drudgery. There was an undertone of anger. Peter and Penelope were not unique and certainly weren't in the midst of a tragedy; they simply were *not* ready. They quickly came to terms with and understood the need for taking a pause and going back to the drawing board. That is when we agreed on a temporary time-out.

The uncomfortable situation they had found themselves in came as a surprise to them both. Prior to the home selection phase, they routinely got along well, had a lot in common, and were very much in sync on a daily basis. All they needed was some chosen time for exploration and evaluation in order to prepare for comfortably navigating this process. Peter and Penelope simply didn't know what they didn't know.

This next assignment is vital to identifying a vivid direction for the home-buying process by learning to compromise while appreciating and celebrating each other's differences as they ultimately relate to planning for the future.

ASSIGNMENT 2

Organize a Strategic Plan

In this exercise, you will need to thoughtfully explore and answer a list of key questions in four categories: Life span, Lifestyle, Location, and Luxuries.

LIFE SPAN: HOW MANY YEARS WILL YOU BE IN YOUR HOME?

- Do you plan to live in your home for three to five years?
- Do you plan to live in your home for five to ten years?
- Do you plan to live in your home for ten-plus years?

Determining the length of time you will own your home is a prerequisite in establishing the necessary parameters that you will need to successfully select the right fit. The duration of owning the home is the amount of time required to receive the present value of future payments, by considering the property's price volatility, resulting from projected changes in the economy.

For example, if you only plan to live in your home for three to five years, you will want to make a very aggressive purchase offer. This offer needs to be in accordance with current market values and based upon the projected market value for the duration of the time you will own the home, while considering that *assumed* values *must* significantly progress in certain areas for short-term ownership, ultimately affecting the equity in your home. Consulting with a properly licensed real estate agent on these values is essential.

LIFESTYLE: HOW WOULD YOU DESCRIBE YOUR LIFESTYLE?

- Active
    - Do you need proximity to greenways?
    - Do you enjoy hiking?
    - Do you need a park for a dog?
    - Do you prefer bike lanes?
- Trendy
    - Do you want ease of access to a social scene?
    - Do you enjoy coffee shops?
    - Do you like shopping areas and boutiques?
    - Do you want walkability to bars and restaurants?
    - Do you need access to cowork spaces?
- Community
    - Are you involved in organizations?
    - Are you a member of a club?
    - Do you volunteer at nonprofit organizations?
    - Are you involved in service projects?
    - Do you prefer to live near your church?

LOCATION: WHAT PART OF TOWN IS MOST IMPORTANT?

- Do you want privacy and serenity, a refuge to get away?
- Do you need a backyard for gardening or a pet?
- Do you need convenient proximity to work?
- Will you start a family there?
- Do you need to be close to specific schools?
- Is building equity your top priority?

LUXURIES: AND PLEASE, ALLOW YOURSELF SOME.

- How many bedrooms do you need?
- How many bathrooms do you need?
- Do you need a garage?
- Do you need storage space for things like bikes, lawn equipment, or holiday decorations?
- Do you need a place for family and friends to stay?
- Do you want an indoor workout area, patio, or pool?

This exercise is designed to be a thorough exploration into wants, needs, and priorities. It may require a lot of soul searching, logistics evaluation, conversations, maybe even a mini-adventure, and hopefully some fun along the way.

One recommendation is to select three different municipalities or neighborhoods in your desired area. Over the course of three weekends, visit each prospective location— one per weekend. If possible, check into a hotel or Airbnb in that area for an overnight stay. Completely immerse yourself in the culture. Make an effort to meet the neighbors who live there. Observe the environment at different times of the day. Learn about seasonal community events, festivals, neighborhood gatherings, and holiday celebrations. Be cognizant of all surroundings—sights, sounds, and smells. Listen for any daily comfort inhibitors such as traffic issues, parking concerns, barking dogs, and trains.

You should come away from this assignment with a sense of *being*. A belonging to a specific energy, culture, and community—and most importantly, a deeper understanding of each other.

Have fun with it!

"Prepare your work outside; get everything ready for yourself in the field, and after that, build your house."

— PROVERBS 24:27 [ESV]

PREPARE

Manage Money and Marketability

Finally, Lake and Lily have gained an understanding of their finances and expenses. They have organized, devised, and committed to a strong strategic plan, and they have determined a lifestyle and potential location that appeals to both of them. Now it's time to contact their favorite (and well-qualified) realtor or real estate agent to begin their house hunting adventure. That's when Lake suddenly has—what he thinks to be—a light-bulb moment.

HONEY, I HAVE A GREAT IDEA!

"Hey honey, I have a great idea—I used to be a winner on my college debate team, and I've negotiated a few capital car deals, and since real estate commissions can be exorbitant, why don't we consider representing ourselves in the real estate transaction? After all, we have all of the latest apps, like Realtor, Trulia, and Zillow. Just think about how much money we could save. Stellar idea, right?"

Wrong.

Although not common, I have encountered the self-appointed real estate *expert* on more than one occasion. It's a strategy I certainly would not recommend.

I recently met one couple who attempted to pursue this concept. We will call them Patrick and Paloma. Patrick was bright and educated and worked in business development for a local company. Clearly, he was a very smart guy. Paloma was a stay-at-home mother to their two young children. I became acquainted with this nice family when they relocated to our neighborhood from Chicago. They had signed a lease on a short-term rental and were looking to purchase their first home. Patrick's strategy was to visit open houses in an attempt to befriend the listing agent who was *representing the seller* of the home. He was naturally charming and personable and was comfortable in conversation. While speaking with the agent, Patrick would subtly devalue other homes in the area that were on the market, in hopes of persuading the agent to reduce the asking price of the listed home. It wasn't long before the agent and other local agents had Patrick's number—and by that, I don't mean his offer number.

What Patrick didn't realize was that agents routinely network with each other. After all, he and Paloma were looking in similar areas and in a specific price range; therefore, it didn't take long before agents caught onto Patrick's *modus operandi*. While Patrick wasn't breaking any laws, his plan of action was flawed. His intention was to save 3 percent of the purchase price by representing himself in buying and negotiating his own real estate transaction. But he completely overlooked a monumental fact: he didn't have an experienced professional with the necessary tools advocating on his behalf. Without these qualifications there was no way of gaining the mountain of pertinent information to ensure the best possible outcome for him and his family.

Real estate agents are more than just friendly men and women who take you to lunch and chauffeur you around in the backseat of their Cadillacs, endlessly discussing home-buying trends and decor ideas. These are highly skilled individuals who have completed hours of educational training, have met the necessary licensing requirements, and have solidified affiliations with reputable brokerage firms. They typically participate in weekly sales meetings and team building opportunities and have active roles in business development organizations and various community associations.

They have exclusive privileges to proprietary resource websites like Multiple Listing Service with access to market insights, and they have firsthand information about pocket listings (homes known to be for sale but not marketed), email blasts, and upcoming listings. They also have the knowledge and experience along with relationships with other agents to develop and produce an accurate comparable of your potential purchase to the market value of other homes in the neighborhood.

In addition to the previously mentioned home-buying strategies, real estate agents have expertise and a support system of other agents and brokers to properly guide you through the contract negotiation process: inspections, property disclosures, title searches, easements, encroachments, appraisals, contingencies, kick-out clauses, and much, much more.

To sum it up, if you are negotiating with the seller's agent, it would be prudent for you to learn the definition of the term fiduciary: "the agent in an agency relationship; receives the trust and confidence of the principal and owes fiduciary duties to that principal." (Dearborn Real Estate Education)

In other words, NOT you. The real estate agent is under legal contract with the seller.

Like Patrick, Lake thought that not having an agent would allow the seller to reduce the price of the home by 3 percent. What he didn't know was that the commission percentage is *predetermined* in the contract between the agent/client, as well as the brokerage firm, and the home owner/seller is legally obligated to uphold the terms of the contract.

BE READY TO MAKE AN OFFER

In order to be prepared to make a strong competitive offer, in addition to becoming your real estate agent's favorite client, you will want to visit your preferred lender to do one of two things: become prequalified for a mortgage or get preapproved. These are two somewhat different distinctions that will establish you as a buyer who has completed the necessary preparation to make a timely and competitive offer.

Prequalification is an estimated investigation into your finances, while preapproval is a more thorough and in-depth discovery of your qualifications as a homebuyer.

This next assignment is a lesson in managing your money and the real estate market by becoming familiar with commonly used terms and gaining basic knowledge of the processes in order to make informed decisions and to properly and collectively manage a path to success.

ASSIGNMENT 3A
Manage the Market

Meet with and choose your "buying" real estate agent to discuss:

- Real estate commissions. Expect to sign a buyer's agreement. It defines a relationship between you and the agent and explains the agent's duties to you.
- Contractual agreement. Real estate agents work on commission.
- Title searches. A property title search is the process of retrieving documents evidencing events in the history of a piece of real property to determine relevant interests.
- Relative comps. These are comparable homes of similar size, condition, age, and style that have recently sold in a certain neighborhood, which becomes the basis for a valid cost comparison.
- Property disclosures. A list of items on a document in which the seller discloses any issues, defects, or previous repairs for the home to inform buyers about any issues with owning the home and to legally protect the sellers from potential lawsuits in the future.
- Experience. Your agent should have recent knowledge through direct observation or participation in the different areas in which you are interested.

ASSIGNMENT 3B
Manage Money

Meet with a preferred lender to discuss loan qualifications. Documents that you will likely be asked to provide may vary depending on the lender and include the following:

- Income:
  - Tax returns. Mortgage lenders typically want to see two previous years of returns to make sure your earnings are consistent with what you are reporting.
  - W-2 forms and pay stubs. Lenders may want to verify current earnings and rule out fluctuations.
  - Other income. If you receive additional income, such as child support, you will need to provide current proper documentation in the form of 1099s, etc.

- Credit history:
  - Credit report. Lenders will want to run a credit report, which will require your authorization. If there are negative credit alerts on your report, be prepared to provide a written letter of explanation.
  - Rental history. Your lender might ask for a reference letter from your current lease holder if you are renting.

- Assets:
  - Investments. 401(k), IRAs, stocks and mutual funds, and other retirement or investment accounts.
  - Bank statements. Lenders may ask for currently active bank statements for risk assessment.
  - Gift letter. If a family member or friend is helping you buy the home, they will need to provide a letter stating their relationship to you, as well as the amount they plan to contribute.

- Debts:
  - Real estate. Do you have mortgages on any other real estate you own or partially own?
  - Credit cards. This information will be on the credit report.
  - Lease agreements. Are you legally obligated on any long-term lease agreements?
  - Divorce agreements. Are you legally obligated to monthly alimony or child support payments?

In addition to the above documentation, you will also need to provide a photo ID and your Social Security card.

DISCUSSIONS

While gathering this information can become tedious and a bit time consuming, lenders perform a careful analysis to determine risk and to provide you with an accurate goal for your success as a borrower.

You will want to thoroughly discuss with your chosen lender current interest rates and the percentage of your down payment. I personally recommend a 20 percent down payment if it's feasible. Many lenders carry products that allow you to put down a lower amount, but in doing so, you will have to carry private mortgage insurance (PMI), which is an added cost. PMI is a way for banks to protect themselves. Lenders will eliminate the PMI insurance when the amount of the loan has been reduced to an 80 percent loan-to-value ratio.

Additionally, do not be afraid to discuss the option of a fifteen-year mortgage. The percentage of each monthly payment increase is minimal compared to the exorbitant amount of interest you end up paying on a thirty-year mortgage over the life of the mortgage—not to mention you will reduce the time it takes to pay off your home completely. And that will be a great day!

"Ninety percent of all millionaires become so through owning real estate. More money has been made in real estate than in all industrial investments combined. The wise young man or wage earner of today invests his money in real estate."

— ANDREW CARNEGIE

PRIORITIZE

Educate, Evaluate, and Edit

Congratulations! It's a great day! Lake and Lily have found their dream home. It's in their favorite neighborhood. It has all of the amenities they were hoping for: proximity to work, three bedrooms, a small backyard, and great bones with potential for an addition. However, there are several cosmetic issues that need to be addressed.

"I can't wait to close on our new home and begin our remodel!" says Lily. "You know, I've never been a fan of dark paint colors—they feel so dreary. I prefer a nice neutral palette. It will brighten up the rooms and create a cool modern vibe."

"Yes, and what about the hardwood floors?" adds Lake. "Yellow shiny stain on oak flooring reminds me of my aunt Betty's cabin, and we must remove that pink tile from the bathroom. When was that in style?"

"Right, and the money we saved on our purchase price

should pay for the remodel. After all, the house cost twenty-five thousand dollars less than we had planned."

Stop right here!

This is a common misconception. While there is no exact science in remodeling, it is imperative that you make an educated guess. I've witnessed many homeowners compartmentalize the different components of buying, remodeling, and selling homes.

PUT ALL OF THE PIECES TOGETHER

When considering a home remodel or complete renovation, there are multiple factors to consider. Remodeling is not just about cosmetics. Your finished product might be beautiful, and you may have done it on an exceptional budget, but if it's not going to sell for what you have invested into it—in the length of time that you will own it—it is not a good deal.

You will need to carefully research several categories, and you will need to make a well-devised list of detailed evaluations before hammering the first nail. This is where all of your hard work comes together—you must put all of the pieces together.

Kitchens and bathrooms are a good place to start, not only for functionality and your own personal enjoyment but also for resale value. However, it is important to remember one general rule. You typically want to avoid making home improvements that will increase your home's market value to more than 20 percent above other comparable home values in your neighborhood.

More importantly, prior to improving aesthetics and function, you will want a thorough home inspection, followed by resolving all necessary items listed in your report. Unresolved safety and building code issues have the potential to become negotiating tools for future buyers. And for some squeamish buyers, these items (most of them easily repairable) could become deal breakers, not to mention they could be hazardous to you and your family.

PICK YOUR BATTLES

This assignment is all about education and evaluation so that you can prioritize and define clear goals to formulate an accurate budget. Controlling costs during this phase begins with revisiting Assignment 1. You will need to review the logistics of your previous strategic plan and begin eliminating any unreasonable and unnecessary costs. Prioritizing all decisions throughout the remodeling phase is your key to success. This exercise truly becomes a lesson in discipline and thoughtful decision-making based on multiple influencing factors, along with carefully and routinely picking your battles in order to protect the return on your investment.

ASSIGNMENT 4
Educate, Evaluate, and Edit

EDUCATE:

- Revisit Assignment 1.
- How long do you plan to be in your home? Discuss with your real estate agent the potential marketability of your home for the duration of the time that you will own it and establish a projected sell price.
- Educate yourself on closing costs, commissions, and any other costs that will be passed along to you. Typically the seller is obligated to certain expenses. Creating a mock closing document will provide you with a good assessment of real estate—related expenses for budgeting your remodel.

Upon accumulating this information, you will have a much needed mathematical equation: the projected future marketability, based on the duration of time that you will own the home, less the closing cost and commissions, and other costs, less all necessary repairs. This will establish your re-model budget.

* Sell price - closing, commissions, taxes, and any other holding costs - repairs = Remodel budget *

Chances are your budget will not be enough to include all of the items on your wish list, but by prioritizing and properly editing unnecessary items, there should be enough in your budget to provide you with a nice finished product.

Next, hire a professional home inspector. Home inspection reports can be quite lengthy. You might even receive a thirty-page spiral-bound notebook, but don't be alarmed. Typically home inspectors are very thorough and will provide vital information regarding the integrity of your home. This report will be full of useful information and facts about your home, providing you with the knowledge to resolve all repairs and maintenance.

Note: I had one buyer attempt to negotiate $5,000 off of the purchase price of a home that I was selling when the items on the report amounted to approximately $500. I opted to take care of those repairs myself.

Now that you have gathered all pertinent details associated with your strategic plan, it is time to make final decisions. You have researched your home's marketability, you have resolved all needed repairs, you have evaluated your areas to remodel, and you have established your budget. You are now ready to move on to the next step.

PRIORITIZE, PRIORITIZE, PRIORITIZE

Your next task is to decide what in your series of items needs to be dealt with according to their order of importance.

Prioritize: "to organize or deal with something according to its priority." (Dictionary.com)

EVALUATE AND EDIT:

- Budgeting. Make a list of the areas you would like to include in your remodel budget. If the estimated amount of the items on your wish list is approximately $25,000 but your budget allows for $15,000, you will need to edit. Careful observation of current home-buying trends may be helpful in guiding you through this step. Kitchens and baths can be good places to start, provided that the roof and windows are in good working order. Replacing faucets, tub and shower fixtures, broken tile, chipped grout, and countertops give an updated look. Paint is a game changer! A new coat on walls and trim give a fresh clean feel. Also consider updating hardware. Outdated cabinet hardware, doorknobs, and old hinges can be an eyesore. Lot cleanup, landscape removal, pressure washing, gutter replacement, and fence repair should always be assessed for the best possible curb appeal.
- Staging. This is a line item routinely checked off of the list. I have worked with many clients who will spend their budget down to the penny, depleting their account and eliminating any opportunity for furniture, rugs, and window treatments. Their project is complete. The walls are painted. The light

fixtures are hung, the tile is laid, the faucets are installed, but there's not one piece of attractive furniture to give their project a finished look. Don't underestimate home furnishings and staging. You can get great deals on classic furnishings online or at consignment sales and local shops. You will enjoy them, the house will show beautifully to potential buyers, and you will move these well-selected pieces to your next fabulous home. So, please allow room in your budget for interior styling.

LET'S GET STARTED. IT'S HAMMER TIME!

Now you are ready to contact a licensed remodeler in your area. I suggest meeting with three contractors to discuss estimating, scheduling, and billing procedures. I also recommend calling up to three references per contractor. You will want to verify licensing and insurance from your contractor. A properly licensed contractor should maintain workers' compensation and liability insurance. Also, discuss who is responsible for obtaining a builder's risk policy.

The items on your list that most likely require permitting are plumbing, electrical, mechanical, and structural. Operating without proper permits can result in faulty workmanship and costly fines. In addition to avoiding unnecessary fines, routine inspections ensure the safety, security, and integrity of your home.

Meticulously discuss the list of items that you have chosen to remodel in your home with your contractor so that he or she can devise an estimate. Don't leave out any detail on your wish list. An experienced contractor should give you allowances for selections such as flooring, tile, hardware, plumbing, and electrical fixtures. Don't be afraid to discuss subcontractor estimates, vendors, and material pricing. Contractors have different ways of structuring their bids; some will use what is called a fixed-bid price, others price by cost plus a percentage of the work, and others may charge a flat fee.

The American Institute of Architects (AIA), and the Associated General Contractors (AGC) have standard construction contracts that provide an excellent resource for reviewing this process.

After beginning construction on your home, hold weekly meetings with your contractor to create open lines of communication and stay up-to-date on progress, as well as discuss scheduling for the coming week. As always, time is money; therefore, consider creating and maintaining a spreadsheet. They are a great way of tracking expenses, disbursements, and percentages of work completed to help you stay on budget throughout the duration of your project.

Now it's time to have some fun, be creative, and see your vision come to life.

"Every act of creation is first an act of destruction."

– PABLO PICASSO

PERSONALIZE

Stage and Style

L ake and Lily are armed and ready with the detailed estimate from their builder for their kitchen and bath remodel. He has given them a quote that itemizes all costs associated with materials and labor, as well as a timeline for completion. In addition, he has formulated a list of specifications that they personally will need to select. Along with this list, he has provided them with cost allowances per line item.

An experienced interior designer, remodeler, or builder thoughtfully devises these carefully calculated ranges and gives them to the homeowner so they can establish a projected target while choosing products and materials. These ranges are intended to give them freedom to make design choices and keep their project on budget, and they may actually provide Lake and Lily with a bit of comfort, as it will narrow down the infinite amount of available options.

Off they go...

A FEW QUICK QUESTIONS

It's Saturday at 11:00 a.m., and Lake and Lily arrive at the home improvement and design center. Their first selection is tile.

"Good morning, Lake and Lily. My name is Glenda, and I will be assisting you in selecting tile for your kitchen and bath renovation. Take a minute to look around and see what you like. I will be happy to devise an estimate and process a work order for you."

"This one is beautiful!" says Lake. "I've always loved white marble. It reminds me of our trip to Europe last year. It was in all of the bathrooms. It's classic and ageless. I've heard it can be fragile, so we'll need to be a bit more careful."

"Yes, I've heard that as well," says Lily. "I don't mind the way marble ages. It gives it a nice patina—you know, a bit of character.

"Great, let's go with this one!"

Glenda returns. "Did you see anything you liked?"

Lake shows Glenda the sample of the white Carrara marble.

"Yes, that is one of our best sellers," she replies. "I will be happy to work up a quote for you. I only have a few quick questions for you and Lily…"

"Will it be going in your kitchen or bathroom?"

"Will it be installed on the floor?"

"Will it be installed on the wall?"

"Will it be going on the backsplash in the kitchen?"

"Would you like twelve-by-twelve squares?"

"Would you like sixteen-by-sixteen squares?"

"How about rectangle?"

"Maybe herringbone?"

"If we go with the square option, do you want it installed in a straight pattern, or in a diagonal pattern?"

"Would you like to pair the diagonal pattern with the herringbone?"

"Would you like to pair the herringbone with the square?"

"Would you like it honed? Or maybe polished?"

"How about grout color?"

"Will there be a tub, and if so, have you selected a tile for the tub surround?"

"Will this be going in the shower, and will there be a seat in the shower?"

"How about a different pattern in the center of the shower floor? A lot of my customers like this!"

"Would you like an accent color on your shower wall? Maybe a pop of color?"

"What about plumbing fixtures? Stainless? Oil-rubbed bronze? Brass?"

"Oh, and by the way, we haven't even talked about kitchen countertops! And I can assist you with appliances tooooo!!!"

QUICK! Call an ambulance!

Well, not really. Lake and Lily didn't suffer any catastrophic health issue—they were just a bit overwhelmed. They were beginning to get a sense of how many more decisions there were to be made.

What should they do? You guessed it—more HOMES work.

This assignment is my personal favorite, but even for the seasoned home remodeler, this phase can become very overwhelming. What begins with a burst of creativity and great enthusiasm can slowly progress into bewilderment and frustration, as there is an endless sea of possibilities when it comes to choosing materials for a home renovation.

CONCEPTUALIZE

We've all heard the saying a picture paints a thousand words, but in this case, it's more like a million words. The glue that visually holds this phase together is conceptualization. Developing and committing to a firm vision is the key to success. Lake and Lily need to pursue each selection with a clear vision of their design concept.

Conceptualization: "an abstract simplified view of some selected part of the world, containing the objects, concepts, and other entities that are presumed of interest for some particular purpose and the relationships between them." (Wikipedia)

More simply said, conceptualization is a formulation of a clear idea.

I occasionally frequent antique malls with a good friend named Tallulah who enjoys collecting unique pieces. Some of us have a passion for shoes; she has a passion for chairs. Tallulah rarely saw a chair she didn't like. With each visit to the antique shop, she saw a new set of one-of-a-kind chairs that, for her, was better than the last set. It was clear that she didn't have a concept. This is easy to do, and I have certainly done it myself. With so many beautiful choices, it is easy to find many things to get excited about, but the key is having a laser focus on the end result. Lake and Lily don't know it yet, but they are about to rehearse their design mantra.

THE MANTRA

Repeat after me, "Does this fit into our concept?"

They must ask themselves this question multiple times throughout this phase. This exercise creates continuity and cohesiveness in style, color, texture, scale, and finishes. This is where vision boards come in handy. Design elements of a room create a mood, and we are all drawn to certain energies and atmospheres. Taking time for self-exploration, determining areas of functionality, and becoming familiar with various design elements will lead to a clear conceptualization.

Developing Pinterest boards are an excellent place to

begin. They are full of wonderful ideas, easy to organize, and effortless to share with your builder, remodeler, or design assistant. Houzz is another great resource for researching home trends, and many products are available for purchase. Many other online retailers will allow you to create a wish list and organize products such as lighting, plumbing fixtures, cabinet hardware, home decor, and more, room by room. This is a great way to group selections as well as track expenses.

Through this practice you may find that you are drawn to a minimalistic modern vibe with clean lines and sleek finishes. Or that you gravitate to a classic traditional style filled with antiques and collectibles. Or a combination of the two, a more transitional look.

You might want to consider cities you have visited and enjoyed and explore the things that drew you there—the sights, sounds, and smells. Was it the architecture or the scenery? Was the atmosphere lively and passionate or quiet and subdued? These explorations will lead you to your personal sense of style and design concept.

ASSIGNMENT 5
Stage and Style

SELF-DISCOVERY:

- What room atmospheres are you drawn to? Do you enjoy the calming effects of the sea or the serenity of the mountains?
- Are you drawn to quiet intimate spaces or rooms filled with light and volume? I have a friend who is drawn to rich jewel tones while another is completely agitated by them.
- What home decor styles are relatable to your area? Certain styles are indicative of their particular region.

FUNCTIONALITY:

- Do you need a work space or an in-home office?
- Do you need a personal space for reading and solitude?
- Do you need a craft area for projects, painting, or pottery?
- Do you need a play area for children?
- Do you need a pet zone for bathing or housing a pet?

DESIGN ELEMENTS:

- What color schemes are you visually drawn to?
- What colors comfort you? (What pleases us visually may not be what comforts us.)
- What textures are you partial to? Do you prefer the light and

airy feel of linen or heavy textures like wool and velvet?

- Do you like things neat and put away or a room filled with accessories and collectibles?
- Do you like clean lines and modern rooms or the graceful details of French pieces?

STYLE CONCEPT:

- Modern and contemporary incorporates reflections of the modern era with current contemporary amenities.
- Classic and traditional assembles elements from a variety of centuries, creating a classic and elegant style.
- Transitional is a contemporary style that mixes traditional and modern styles.
- Coastal incorporates reflections of ocean hues and sun-bleached fabric colors.
- Eclectic and artist present a mix of trends, periods, and colors.

RESEARCH:

- Websites. View recent renovations and current trends.
- Housing apps. Create vision boards.
- Magazines. Gain an understanding of different design styles.
- Furniture stores. Explore current trends.
- Home improvements centers. Become familiar with available options.

Exploring your passions and interests launches an emersion into different cultures, history, geography, societies, literature, music, and more. Through these adventures, explorations, and discoveries, you will develop knowledge and curate a collection of items that will ultimately define your personal style and artistry.

This is *your* home!

"There is no value in life except what you choose to place upon it, and no happiness in any place except what you bring to it yourself."

— HENRY DAVID THOREAU

PROSPER
Lake and Lily Live Happily Ever After

My hope is that the information in this book has provided you with a basis of knowledge to begin your own wild and exciting ride into home development, along with a personal blueprint for success in whatever phase you are considering. Home-buying, renovating, and selling is a complex process of analytics, thoughts, and emotions, making it a very personalized journey.

Your home is the nucleus of your universe—the alpha and the omega of each day.

With each project, I have gained a greater awareness of the seven spheres of society and how they relate to our homes. According to celebrity life coach Mike Bayer, these spheres are social, personal, health, education, relationship, employment, and spiritual. We experience each of these influencers each day. Our outside environments send us constant subliminal messages so that we never realize that our atmospheres are being shaped by messages out of our control. Being cognizant is the first step in gaining power over our own surroundings.

Our home encapsulates us by encompassing the seven universal spheres of our own well-crafted environment. We are free to carefully, willingly, and thoughtfully curate an atmosphere of influencers that will provide us with our own personal climate for growth and comfort.

How will you create yours?

Social. Your home is a place of fellowship: a place for celebrating milestones, recognizing accomplishments, making cherished memories of special occasions, and hosting friends and family.

Personal. Your home is a place of privacy where you relax and unwind.

Health. Your home is a place of comfort where you recuperate from illness or injury, replenish and nourish physical needs, rest your mind, and calm emotions.

Education. Your home is a place of study and enlightenment through reading, research, music, art, and more.

Relationships. Your home is a place to connect and cultivate love, laughter, and a lifelong foundation with family.

Employment. Your home is a place of refuge for rejuvenating from a day in the professional world.

Spiritual. Your home is a place of meditation, reflection, prayer, giving thanks, and replenishing the human spirit.

Through self-exploration, you examine and realize your

individual spirit, your own intellectual interests, and your sense of humor. As you explore your passions, you cultivate your personal style and love of learning.

Your home is your Picasso, your Monet. Your own unique work of art.

There are no judge or jury.

It should represent all that YOU are.

"Little pink houses for you and me."

– JOHN MELLENCAMP

AFTERWORD

I've had yet another moment of self-discovery through writing this book. If you had told me as an adolescent, teenager, or young adult that I would find myself involved in the construction world, I wouldn't have believed you, but here I am.

After more than twenty years of experience being involved in more than a hundred projects—rehabbing and living in fifteen of my own homes—I guess you could say that I have learned a few things.

Being an artist by nature, I continue to find endless joy in the creative process; however, I have also developed a great appreciation for the logistics of home development, and I understand the absolute necessity of "Putting All of the Pieces Together."

I have been part of a variety of projects including new construction, renovations, and remodels, as well as commercial. Building and development have given me the opportunity to learn the tactical application of building, from footings and framing, to banking and finance, to careful planning, estimating, and budgeting, followed by marketing and selling.

Home development has challenged my physical capabilities, as purging, packing, moving, organizing, and setting up a home for a family can require great stamina. It has also challenged my cognitive skills required for estimating, budgeting, scheduling, and planning.

All of this has provided me with a rich education as a business person and as a curator, along with an ongoing love of learning for both residential and commercial real estate growth and development in my hometown of Nashville, Tennessee. This journey has introduced me to a variety of lifestyles, beginning with a farmhouse surrounded by cattle, followed by a beautiful Mediterranean-style home in the suburbs, to a sprawling ranch with a backyard pool, a mid-century modern, to a taste of urban living with a view of the city. While always keeping an eye on the next project, I've had moments of sadness in saying goodbye to the last chapter.

In hindsight, the writing was on the wall, or should I say drywall, that went unrecognized. As a young child, I typically had a nightly sketch project. My mother and sister found great amusement in waking up one morning to learn that the night before I had drawn a fairly accurate sketch of the kitchen elevations (from a nine-year-old perspective, of course). Another might have been that my favorite book as a preschooler, which

I had my grandmother read to me over and over, was *The House That Jack Built*. I suppose that's what you might call divine foreshadowing.

Through the growth and development of curating a home comes growth and development of ourselves—and vice versa. There are a lot of opportunities in the world, a lot of places to explore, and a lot of new people to meet.

What's your next project?

HOMES WORK

ASSIGNMENT 1

Have a Clear Understanding of Your Expenses

MONTHLY DEBT:
- apartment rent | ___
- car payments | ___
- student loans | ___
- credit cards | ___
- other legal obligations | ___

YEARLY TAX OBLIGATIONS:
- income tax | ___
- franchise and excise tax | ___
- schedules C or S | ___ (Consult with your CPA to gain an understanding of a potential yearly amount.)

UTILITIES:
- gas | ___
- electric | ___
- water | ___
- garbage | ___
- Wi-Fi/cable/streaming subscriptions | ___
- cell phones | ___

INSURANCE:
- renter's insurance | ___
- auto insurance | ___
- health insurance | ___
- life insurance | ___

MEDICAL:
- co-pays | ___
- doctor visits | ___
- prescriptions | ___

AUTO:
- gasoline | ___
- oil changes | ___
- maintenance | ___

FOOD, BEVERAGE, AND ENTERTAINMENT:
- dining out | ___
- annual vacations | ___
- family visits | ___
- unreimbursed work-related travel | ___

CHARITABLE CONTRIBUTIONS:
- church tithing | ___
- nonprofit donations | ___

INCIDENTALS:

- haircuts | ___
- mani/pedis | ___
- gym memberships | ___
- gifts | ___
- cash | ___

ASSIGNMENT 2
Organize a Strategic Plan

Remember: Consulting with a properly licensed real estate agent is essential.

LIFE SPAN: HOW MANY YEARS WILL YOU BE IN YOUR HOME?
- Do you plan to live in your home for three to five years?
- Do you plan to live in your home for five to ten years?
- Do you plan to live in your home for ten-plus years?

LIFESTYLE: HOW WOULD YOU DESCRIBE YOUR LIFESTYLE?
- Active
  - Do you need proximity to greenways?
  - Do you enjoy hiking?
  - Do you need a park for a dog?
  - Do you prefer bike lanes?
- Trendy
  - Do you want ease of access to a social scene?
  - Do you enjoy coffee shops?
  - Do you like shopping areas and boutiques?
  - Do you want walkability to bars and restaurants?
  - Do you need access to cowork spaces?

- Community
  - Are you involved in organizations?
  - Are you a member of a club?
  - Do you volunteer at nonprofit organizations?
  - Are you involved in service projects?
  - Do you prefer to live near your church?

LOCATION: WHAT PART OF TOWN IS MOST IMPORTANT?

- Do you want privacy and serenity, a refuge to get away?
- Do you need a backyard for gardening or a pet?
- Do you need convenient proximity to work?
- Will you start a family there?
- Do you need to be close to specific schools?
- Is building equity your top priority?

LUXURIES: AND PLEASE, ALLOW YOURSELF SOME.

- How many bedrooms do you need?
- How many bathrooms do you need?
- Do you need a garage?
- Do you need storage space for things like bikes, lawn equipment, or holiday decorations?
- Do you need a place for family and friends to stay?
- Do you want an indoor workout area, patio, or pool?

ASSIGNMENT 3A
Manage Marketability

MEET WITH AND CHOOSE YOUR "BUYING" REAL ESTATE
AGENT TO DISCUSS:

- Real estate commissions. Expect to sign a buyer's agreement.
  It defines a relationship between you and the agent and
  explains the agent's duties to you.
- Contractual agreement. Real estate agents work on
  commission.
- Title searches. A property title search is the process of
  retrieving documents evidencing events in the history of a
  piece of real property to determine relevant interests.
- Relative comps. These are comparable homes of similar size,
  condition, age, and style that have recently sold in a certain
  neighborhood, which becomes the basis for a valid cost
  comparison.
- Property disclosures. A list of items on a document in which the
  seller discloses any issues, defects, or previous repairs for the home
  to inform buyers about any issues with owning the home and to
  legally protect the sellers from potential lawsuits in the future.
- Experience. Your agent should have recent knowledge
  through direct observation or participation in the different
  areas in which you are interested.

ASSIGNMENT 3B
Manage Money

Meet with a preferred lender to discuss loan qualifications. Certain documentation requirements may vary depending on the lender.

INCOME:
- Tax returns. Mortgage lenders typically want to see two previous years of returns to make sure your earnings are consistent with what you are reporting.
- W-2 forms and pay stubs. Lenders may want to verify current earnings and rule out fluctuations.
- Other income. If you receive additional income, such as child support, you will need to provide current proper documentation in the form of 1099s, etc.

CREDIT HISTORY:
- Credit report. Lenders will want to run a credit report, which will require your authorization. If there are negative credit alerts on your report, be prepared to provide a written letter of explanation.
- Rental history. Your lender might ask for a reference letter from your current lease holder if you are renting.

ASSETS:

- Investments. 401(k), IRAs, stocks and mutual funds, and other retirement or investment accounts.
- Bank statements. Lenders may ask for currently active bank statements for risk assessment.
- Gift letter. If a family member or friend is helping you buy the home, they will need to provide a letter stating their relationship to you, as well as the amount they plan to contribute.

DEBTS:

- Real estate. Do you have mortgages on any other real estate you own or partially own?
- Credit card. This information will be made available on the credit report.
- Lease agreements. Are you legally obligated on other long-term lease agreements?
- Divorce agreements. Are you legally obligated to monthly alimony or child support payments?

OTHER DOCUMENTS:

- Social Security card
- Photo ID

ASSIGNMENT 4

Educate, Evaluate, and Edit

Make a list of the areas you would like to include in your remodel, and then educate, evaluate, and edit. Remember:

*sell price — closing, commissions, other costs
— repairs = remodel budget.*

KITCHEN:

- appliances
- cabinets
- countertops
- knobs/pulls
- flooring
- tile

BATHROOM:

- plumbing fixtures
- cabinets
- countertops
- knobs/pulls
- flooring
- tile
- mirror

OTHER AREAS:

- floor refinishing
- doors
- framing
- trim
- shoe molding
- hardware
- lot cleanup
- landscape
- tree removal
- driveway

STAGING:

- furnishings
- draperies
- blinds
- rugs
- lamps
- accessories

EVALUATING A BUILDER/CONTRACTOR/REMODELER:

- licensing
- estimating
- scheduling
- billing procedures

- liability insurance
- workers' compensation
- insurances that are your responsibilities
- permitting
- references

ASSIGNMENT 5

Stage and Style

SELF-DISCOVERY:

- What room atmospheres are you drawn to? Do you enjoy the calming effects of the sea or the serenity of the mountains?
- Are you drawn to quiet intimate spaces or rooms filled with light and volume? I have a friend who is drawn to rich jewel tones while another is completely agitated by them.
- What home decor styles are relatable to your area? Certain styles are indicative of their particular region.

FUNCTIONALITY:

- Do you need a work space or an in-home office?
- Do you need a personal space for reading and solitude?
- Do you need a craft area for projects, painting, or pottery?
- Do you need a play area for children?
- Do you need a pet zone for bathing or housing a pet?

DESIGN ELEMENTS:

- What color schemes are you visually drawn to?
- What colors comfort you? (What pleases us visually may not be what comforts us.)
- What textures are you partial to? Do you prefer the light and

airy feel of linen or heavy textures like wool and velvet?

- Do you like things neat and put away or a room filled with antiques and collectibles?
- Do you like clean lines and modern rooms or the graceful details of French pieces?

STYLE CONCEPT:

- Modern and contemporary incorporates reflections of the modern era with current contemporary amenities.
- Classic and traditional assembles elements from a variety of centuries, creating a classic and elegant style.
- Transitional is a contemporary style that mixes traditional and modern styles.
- Coastal incorporates reflections of ocean hues and sun-bleached fabric colors.
- Eclectic and artist present a mix of trends, periods, and colors.

RESEARCH:

- Websites. View recent renovations and current trends.
- Housing apps. Create vision boards.
- Magazines. Gain an understanding of different design styles.
- Furniture stores. Explore current trends.
- Home improvements centers. Become familiar with available options.

ABOUT THE AUTHOR

ABBE QUARLES is a Nashville native, contractor, and former owner of a local residential construction company. In her twenty-plus years of experience in real estate and home improvement, she has been part of more than a hundred projects, including kitchen and bath remodels, complete home renovations, and luxury home construction. Her former company was voted one of Nashville's Future 50 Companies by the Chamber of Commerce and was featured in the *Nashville Business Journal*, *Nashville Interiors Magazine*, and other local publications.

Throughout her years of experience, Abbe has developed relationships with skilled craftsman and reputable vendors who provide her with firsthand information regarding local construction industry standards. Abbe has completed projects from mid-century modern and classic Georgian to urban contemporary, allowing her to appreciate each client's individual style.

She currently owns AQ Collective LLC, a residential

development consulting group that focuses on establishing personalized goals for each client. Their planning process includes home-buying, renovating, and selling by putting all of the pieces together. Abbe has completed the educational requirements and examination for the Tennessee Board for Licensing Contractors, and she and her team have met the educational and licensing requirements for the Greater Nashville Association of Realtors, providing exclusive access to resource proprietary websites such as the Multiple Listing Service, giving them knowledge of the market, pocket listings, email blasts, and new developments.

9 781734 355604